Reflections on Five Decades
Electrifying the Built Environment

by Joe H. McFatter, Jr.

Copyright 2018

Preface

Recently a major event in my life offered the blessing of an epiphany: after some reflection, I came to grasp that it was time to bring my fulltime engineering career to a close. The choice was sort of "forced" on me, and I had been fighting off that decision for a while, wanting to find a better point to begin my "retirement" (I do not relish that word as so many do.) But the combination of experiencing another "reduction in force", coupled with realization that aging was beginning to take my energy, along with an increasing tremble in my right hand (I am right handed), and one other aging issue, showed me that it was time to stop working fulltime – get off the race track – and turn to a new chapter in my career and life. For some months since I was laid off, I have spent a lot of time soaking in what retirement feels like, and in my case, I can see I need to remain active in some capacity, not being a golfer nor having other hobbies to hold my fascination and continue to hone my mind. While I have a passion for music, and do play percussion (congas), I cannot deny my basic nature: I was born to be an engineer, and in some capacity I need to continue to work that basic aspect of my being.

So I have a lot of time on my hands presently, and I thought well, it might be interesting to some engineers, aspiring engineers and perhaps others, to hear my own personal reflections as a professional electrical engineer. From this writing effort may also come another personal epiphany, perhaps a catharsis that brings light to the path before me. For now, I have decided to start a one-man part-time consulting business, not doing engineering, but drawing on the spectrum of my engineering background and life experiences.

All of us have a personal life that we bring to our work lives, at the office, or job site, or where ever. It is my intent to keep this book focused on the "engineering", and for it not to be a personal memoir or biographical resume. Suffice it to say that as a man of European descent who married an African-American beauty, working in the mostly "white" world of typically "conservative" engineers, there were challenges over a large part of my career. Fortunately the world turns and workplace civility continues to improve. The purpose of the book is to speak to the types of engineering projects that I have been involved with, what made those challenging and at times very unique, and reflect on how the design process itself has evolved over these decades. Out of this review, I would hope that by sharing my experiences a prospective young student of engineering may find a few nuggets that could help them making a key decision about their education and future. Mainly for my own purpose, I have organized the book into chapters representing segments of my career.

Introduction

About a year after the Soviet's put Sputnik I into orbit, my parents decided to give up their ranching heritage life, that was a continuing financial struggle, and move our family to southern California. We landed in San Diego, in the Claremont Mesa area, where my folks bought their first house, and my father found a good job as an electronics quality control inspector. I was twelve years of age, a pure country boy, uprooted from a tiny town school and suddenly just one more student in a junior high school (7th-9th grades). Not long after our move, John F. Kennedy announced America's race to the moon, which would consume that decade. I had always been a pretty smart student, always inquisitive as well, and soon found that this new school offered a couple of classes that would prick my potential to one day be an engineer. I took drafting classes and electrical shop classes, the latter for all three years. In electrical shop we actually had a book that was far beyond our normal level, and we would get some theory then we would go out into the shop and actually build projects from scratch. I remember building a small DC motor, including the wood work, winding the rotor with copper wire, making the commutator; and, it worked, much to my delight. Then there was a silicon diode radio, which also worked, and using headphones I could listen to a couple of local stations. In the ninth grade I tackled building an amplifier from elemental components (diodes, resistors, capacitors, inductors) that my dad brought home for me to play around with, from the General Dynamics plant salvage yard. So, although I didn't announce it outright, in my heart I knew by the age of fifteen what I wanted to study in college, and put my mind to taking courses that would lead me to that goal.

I began my career in electrical engineering after a stint as a lieutenant in the United States Air Force, during the Viet Nam war era. In the spring of 1969 I graduated with a bachelor's degree in electrical engineering, and having been in ROTC for four years, headed off to USAF pilot training. Growing up I had visions of getting my engineering degree, then becoming an Air Force pilot, then becoming an astronaut. But, life being the journey it is, I dropped out of pilot training after completing the primary jet training, because it was clear to me, even though I was at the middle of my class ranking, that flying was just not what I wanted to really do after all. I give thanks even today that I was not sent straight to Viet Nam. Rather I was sent to another school in Illinois, then assigned to the nuclear missile squadron at Minot AFB, North Dakota, where I spent two years as a junior officer in a couple of roles. Then along came the opportunity to leave active duty early, and I and others jumped at the chance. I had become totally disillusioned with the military life, the inequities between the commissioned and enlisted, the biased justice system, and frankly the circus that the Strategic Air Command, at least on the missile side, had become.

I left the thawing plains in April 1972, not looking back, and headed south in my Opel GT, burning up the highways in my haste. Returning to Austin, Texas, I frankly was clueless on what I wanted to do, or be, for that matter. I had come back to Austin, where I attended The University of Texas, to be with the beautiful woman, also a UT graduate, who I had wed a few months before. It wasn't until I arrived and settled in with my bride, who did have a job, that I really needed to get on the ball and find one myself. We had toyed with the idea of moving to California, where I might go to graduate school, but the exigencies of providing for our livelihood quickly evaporated that notion.

Oddly, while I was still on active duty at Minot Air Force Base (Strategic Air Command), I found myself with some extra time on my hands, and on finding out that I had an electrical engineering degree, the base civil engineering group asked my commander if I could lend them a hand working on some simple engineering projects, such as designing power to motor vehicle heater receptacles. Anyone who has lived in that northern clime knows what this refers to, but: cars, trucks, and other vehicles parked outdoors must be "plugged in" when not running, to power the water tank and oil heaters that keep the engine from freezing during the long and brutally cold winter nights. I jumped at the chance to do this, although frankly I knew very little about doing the design, but was given some original drawings to use as a guideline. Out of this experience I realized that I should think about one day working as a professional engineer, a "PE", or at least having that certification. I had also been taking an electrical engineering correspondence course (this was long before we would begin to do such courses "on line"), as a refresher to what I had studied at UT. I guess my engineer blood began to really percolate, because I decided to take the first written examination that is required of professional engineers. This I did, and having done well on the exam, gave me the signal that a career in engineering within the built environment might just well be what I needed to pursue. So I did in fact return to Austin having a jump start.

Always being the optimist, after a few days I also began to look around Austin for a house that we might buy or rent, as we were in an apartment complex, and in my poking around and calling I met this lady realtor. Sharing my story and predicament of not having a job with her, and that I had an electrical engineering degree, caused her to smile and inform me that her own son was an electrical engineer, and that I should see him about employment. As I recall she let him know to expect my call.

After an interview with him, in a small engineering office of some five men in downtown Austin, I started work the following Monday. Thus began my long and rewarding career as an electrical engineer designing power and related systems for the built environment.

If the reader is unfamiliar with the term, "built environment", here is one definition:

"In social science, the term **built environment**, or **built** world, refers to the human-made surroundings that provide the setting for human activity, ranging in scale from buildings to parks. It has been defined as "the human-made space in which people live, work, and recreate on a day-to-day basis." (Wikipedia)

I will take license in relating my own career, to expand this definition to include the generation and distribution of electrical energy to those places of human endeavor, recreation and commerce.

Chapter One – "In Training"

So there I was, my first day on the job as a "civilian" (I was still a commissioned officer on reserve status). Sitting at my new "desk", an ancient wooden drafting table, on a stool, suddenly I felt like an engineer! Had I only known how far from reality that feeling was, I might have quit at the end of that same day. I suppose all engineers beginning their careers feel that way, as probably do others in other professions. In the early 1970's engineering design was still firmly planted in what today could be called the "dark ages". Like all the other engineers and designers in the one production room where I sat, we all worked at drafting tables with parallel bars. Our office did not use what generically was called the universal drafting machine, and such apparatus really were just too cumbersome for the electrical, HVAC and plumbing design that the firm did for architectural projects. For someone unfamiliar with this type of endeavor, this type of consulting firm, and industry that still flourishes today, "puts" the "guts" into a building design: designing the heating-ventilation-air conditioning systems, electrical lighting and power (and other electrified systems), and plumbing (and fire protection systems). Each of these systems have many special symbols to represent specific elements and components, interconnected graphically on plans to depict the intended design.

I should mention that the arsenal of tools I had at my disposal, typical then for that type of design, were a couple of protractors, a triangular architectural scale and similar engineer scale, several plastic triangles and special shape devices, ink pens, and the electric eraser (which I became very proficient at using). I also had, though seldom had to use, my nice set of drafting compasses from my college drafting course. I should mention that during that era, to my knowledge, probably most engineering students took a basic drafting course. Frankly I got a C in the course, and probably didn't deserve that. My biggest challenge was hand lettering, which I never did master.

Also, I had at my disposal my old college slide rule. I think I might have used it only a few times at the start of my career, as by the early 1970's the electronic calculators begin to take a foothold in the industry. However, these did not exist during my college days, and every engineering student could be spotted by the big slide rule often carried hanging from the young man's belt. (I do not remember how females engineering students carried theirs, oddly; there were so few female students. Back then students did not use backpacks that became common years later: we all just carried our books under our arms, or in a satchel usually.

However, most of what I began drawing that first day, and daily for months, were orthogonal lines, often in a 2-ft x 4-ft grid pattern, representing ceiling tiles that were typical for office spaces at that time (and for decades later); therefore, a fancy T-bar was not even necessary, just an old fashioned set of plastic triangles aligned on the parallel bar. My assigned task for the first few weeks was not that technically difficult: performing basic photometric calculations using what is called the "zonal cavity method", and selecting lighting fixtures (usually "2 x 4" fixtures having fluorescent lamps) to meet the recommended illumination levels for the given work spaces in the design plan. Having selected the fixture, I then would complete the grid pattern on the architectural plan drawing, adjusting the grid as needed to align with necessary reference points, and then draw in the outline of the lighting fixtures, positioned on the spacing pattern that would come closest to providing the calculated illumination level.

At that stage in the annals of designing architectural projects electrical systems (and mechanical and plumbing, also), the typical method was using black ink on white drafting paper. As I recall it was a good two weeks of continuous drafting before I could attain an acceptable product. Ink smears and mistakes in line drawing or other symbols required the trusty electric erasure, worked in conjunction with erasing shield (a flat, thin piece of stainless steel with various shapes punched in it). I probably became a much better eraser than drawer, really!

An electrical design for buildings uses graphical symbols, each associated with a descriptive legend. For instance, 120volt wall receptacles of the duplex type (capable of receiving two plugs as is typical in offices, or residences), has a special symbol. Specific symbols for each component of an electrical system are used. So the first few months I was getting familiar with these symbols, placing these on various sheets of plan view "2-D" drawings for a given project, interpreting requirements based on the type of project and the intended use of each space, or room, or even outdoors. I learned all this by looking at similar projects the firm had produced, but moreover, beginning to study and learn the National Electrical Code (NFPA 70).

That aspect of electrical power design that was more "engineering" than "drafting", was creating what is called the single-line diagram, or the corollary drawing called a "riser diagram". Depending on the project, either or both may be elected to be drawn. The riser diagram is more pictorial of actual layouts and routing of interconnecting major circuits, called "feeders", as well as what is called the "service", which is the interconnection to the serving electrical utility company's system. Developing the design for what these depictions show, requires calculations following, as a minimum requirement, the rules within the NEC, as well as abiding by any overriding municipal or state ordinances.

After some nine months of heads down design and drafting, one day I was called into the office of the newly arrived, but grizzled office manager, and informed I was being let go. I was caught off guard, but being the young and fearless guy I was (still fearless, mind you, but a lot more cautious), I merely went to one of the competing firms across town, and started a new job the next Monday at a nice salary increase, plus the promise of lots of overtime. They were not kidding about the overtime either, as I often was working at least 50, sometimes 70 hours per week, and the additional income was a blessing to my new bride and I, allowing us to move to a tiny single-family house - which we promptly decorated - within walking distance of the office.

This small firm had a different "model" of production than the prior firm: they had three principal owners, each with a segment of responsibility basically aligned with the practice discipline (mechanical, electrical) of each principal. Working more or less directly under the principals were designers and "EIT's" (engineers-in-training) like myself, who did virtually all the design of MEP systems, oddly doing what amounted to drafting, but using red pencils! These red-lined design drawings were then given over to the drafting department, who were in effect "tracers", given that they would place a red-lined drawing beneath a sheet of vellum paper, and literally trace what had been drawn by the designer, but greatly improving on the quality of the final drawing and the hand lettered notes. Looking back on this system over my years I thought this an odd way to produce design drawings, but at that time, for that firm, it made good sense. In any case, I avoided having to do real drafting, and yet fully control how my drawing content would appear. We also had my first exposure to quality control, in that when the drafters completed a drawing, a print was made, and returned to myself and the other designers along with our red-lined drawing, for checking and further development as needed. One aspect of this system was in fact that as the designer I could bring my design and the appearance of the drawing composition to completion, or nearly so, before it went to the drafters, which cut down on iteration after iteration of design-draw-correct/revise that I saw later in my career. So it actually was a pretty good production method at the time.

I was fortunate at this firm in also being given nearly full responsibility for assigned projects, with very little supervision. While I could go to a principal to ask about this or that, the fact was, that I probably knew more about electrical design than anyone else there, so they just tossed me in the water, and I was determined to swim. I had always had a curious and fairly studious disposition, so not having a mentor or tutor, I had to seek out the knowledge needed to perform my design. I was forced very quickly to dig deep into the National Electrical Code, and to further research technical topics on electrical equipment and systems, by delving into engineering literature produced by major electrical manufacturers, such as General Electric and Westinghouse, whether the topic be illumination or some type of switchboard.

One project I was thrown into was a major expansion to the City's Hospital, adding a nursing tower to the old and sadly poorly maintained Breckenridge Hospital. I vividly recall one site visit to gather information of the existing electrical system, and seeing dirt-dabber wasps nests and spider webs inside what constituted essential electrical panels (and also hearing stories of rust falling from air conditioning diffusers in OR!). This project required my full immersion in research, talking to all kinds of systems technical and sales representatives, to get my mind around various requirements such as the emergency power system, and the myriad of signal and communications systems that were totally new to me. Somehow I waded through all this new information and was able to create the necessary designs of the lighting, power and "low voltage" systems. For decades I carried with me a set of prints of this first big project in my career, and over those same years that hospital has gone through many changes and expansions, no doubt replacing virtually everything I had designed.

Another project that was dropped on my design table was the rebuilding of overhead power distribution and services to a public housing project. This again was my first, and I knew to be successful in accomplishing the design, I had to study all I could find about such design. The project involved wooden pole-line construction, pole-mounted transformers to step the voltage down, and aerial service drops from these poles to the buildings. I turned to the resources I could find, including a Rural Electric Cooperative (REA) design manual which had all kinds of details and hardware information, and my Handbook for Electrical Engineers, and literature from manufacturers like General Electric, as well as others for other components such as lightning arrestors, guy lines, insulators and hardware. With this material in hand I spent hours studying while actually pulling the design together. Since this was a renovation project, it entailed removing the entire existing system and replacing it, therefore I spent a few days at the housing site taking photos, sketching and making notes. After a few weeks, working with my drafters, we completed the design, and I thumped my chest you can bet!

Another "odd ball" project soon arrived on my desk. I suppose after they saw how I tackled the public housing job, they could count on me to dig into this industrial project. Near Austin at that time Alcoa Aluminum had an aluminum plant, which involved of course smelting, by means of electric furnaces. One of the large, high voltage "rectifier transformers" required replacement of the output "bus bar" connectors from the transformer to the rectifiers. So my task was to design the replacement. Since I knew nothing at all about this, I latched onto the plant electrical engineer who had it basically figured out, and just needed an engineer to do the hard design for contract bidding. I learned something about high voltage transformers and bussing, that would stay with me my entire career. Any young engineer would benefit from approaching any given project from a greater holistic perspective, and attempt to learn not just the focal details, but also embrace the layers of surrounding process, whatever it may be.

It became clear to me that I could do more than what I was tasked at this firm, and given the low salary situation, I knew it was time to move on and be able to provide better for us. So I begin to search for jobs with large firms or industrial situations, and took a couple of interviews with large firms, but for whatever reasons was not hired. Then I saw an ad placed by Dallas Power & Light Company, and scored an interview with their head of power plant engineering. Their offices were in the DP&L building on Commerce Street, and after working for small firms in Austin, I was duly impressed with the surroundings and the opportunity to "step up my game". Fortunately I was hired, and my wife and I loaded a truck with belongings and our several cats, and headed for Dallas, finding a cute cottage to rent in the Oak Cliff area of Dallas (which we later were able to buy on my Veterans benefits).

Chapter Two – "Getting My Stamp"

My new job in Dallas proved to be just perfect for furthering my knowledge of the broad scope of industrial power and controls design. I was assigned to work directly for a very senior engineer who had "grown up" as an engineer within General Electric, then left to work in the electric utility business. Ed was a quiet, thoughtful and studious mentor, who also was meticulous and required completion and accuracy. He also was a great tutor and mentor.

My assignments for the first year or so were diverse, and mostly "plant betterment" activities which allowed me to use similar existing designs, specifications and documentation to execute the new work. Sometimes it might be replacing a number of pressure or flow metering devices, and in that case I would make a trip from the downtown office out to the particular generating plant station. I should mention that at that time the Company had six stations scattered around the Dallas area, and each station had multiple generating plants, ranging up to 750 megawatts of output, using gas-fired steam boilers, including what is called "super critical" type. These plants had been constructed over the prior roughly 75 years or longer. (One, located virtually right near downtown – where now stands the large sports arena – was equivalent to a museum it was so technically antiquated.)

On other occasions I might be assigned the task of preparing documentation for replacing a 138kV oil-filled circuit breaker in a generating station switchyard. All these activities gave me the vantage of learning not just about the specific device or equipment, but how these worked in the greater scheme of the systems, and my always-hungry mind wrapped around these opportunities, filling my knowledge base that served me over decades to come.

Being an established utility company, there were also ample times where we were not actually working on a particular task, thus giving me lots of study time, to pour over existing plant system drawings, equipment books, and to walk the stations absorbing what I observed. I loved going to the plants, although I wasn't keen on actually working "in the field", preferring the niceties of the home office with its huge lunch break room and the benefit of working in proximity to the other engineering groups such as protective relaying, transmission line, and substation design sections. I eagerly sought out conversations over coffee or just walking across the room or hallway to talk to other engineers and technicians, picking their brains on various topics. I frankly never experienced a situation equivalent to this setting over my future career breadth. However, in all my future situations I always sought out those who knew things that I did not, and I strongly urge any young engineer to cultivate the inquisitive mind regardless of the workplace or environment. Engineers regardless of their age or circumstance should always be open to learning from others about matters which do not necessarily impact one's immediate work, but either relate to that work, or might somewhere "down the road". Store your acorn young friends!

Another wonderful aspect about my four years at the "light company", was that I was literally my own boss, meaning I had to be highly self-motivated getting assignments done on time and correctly. This translated to be attuned to the resources available, such as the veteran electricians that actually maintained each plant. Each plant had a maintenance group of electrical, controls and mechanical staff, and most of these guys had been working at their plants for many years – some even when the plant was built, and they knew every nook and cranny of their back yards, so to speak. I leaned on these guys heavily, got to know the leads and "supers", and did my best to win their confidence since to them I was still wet behind the ears. I found, and continued to find over decades, that most workers at sites of any sort are usually proud of their domains, and eager to show off their wares and share their ideas with some "new kid on the block" once the ice is broken. I have always approached such people who I knew to hold the key to my ultimate success, with deferential respect and often a bit of "good ol' boy" joshing, so long as it didn't go too far. It's an approach I recommend. If you show someone empathy, that also helps them get over any barriers to true communications.

Reflecting on some projects that I did "from scratch", two come to mind: one consisted of our team designing waste water treatment facilities at all the stations that relied on cooling lakes, and the other was designing a pre-light ignition safety system for an older boiler. The waste water treatment projects came about due to a mandate for operators to start cleaning up their waste water before discharging effluents into the lakes, which all connected to rivers. I worked with two of my peers, both very good young engineers of my age group, one a mechanical who also undertook the process aspects, and a civil engineer who designed the catch basins. I was responsible for realizing the controls systems, power distribution, cathodic protection and grounding of these installations, each unique to the station being served. We not only designed the systems, but actually were responsible for the selection of each equipment and device, preparing detailed construction drawings, and monitoring the work in the field. I learned a lot about controls during this project. Regarding the pre-light safety system, one of the plants had a boiler that did not have such a system, but such a safety was needed to assure that the sequence of ignition did not allow a condition where gas had accumulated prior to firing the igniters. I had to really study all the elements of this boiler lighting operation, and work with the plant operators to come up with a system that would work and was "GI proof". Over the years I always enjoyed doing one of a kind projects, and fortunately I was often able to get my hands and brain around the more fun and challenging ones. I have found that seeking out the difficult projects over time proved to be the best learning laboratory, and I certainly advocate all engineers to not be shy about undertaking challenges. Whether its sports or engineering, the long term winners will be those willing to take on the tough ten yards head on.

One attribute of power generation that stuck with me and prepared me for later project challenges, was the opportunity to work on and around large industrial equipment and systems, hot, noisy, and potentially dangerous. I had been able to do work assignments on high voltage, heavy power medium voltage (such as designing a portable switchgear and bussing system for emergency use at the Dallas Steam Electric Station, should a major accident occur there that could cut off power to half of downtown Dallas), and to witness what could happen when things go wrong. To wit, one day two mechanics were working to take the stem out of a major steam line valve. The unit had been shut down for some long time, so no one gave thought to what could happen. The worse did happen, as they turned a big wrench, trapped steam energy blew upward shooting the valve stem like a missile impaling one mechanic and carrying him over the side of the turbine deck to the ground far below. Another major incident was when a 750MW turbine over-speed due to a protective relay failure, and the generator totally flew apart internally. I was able to see it, and the inside of this huge machine looked like scramble eggs. Over the years happenings like this, related to design failure and safety systems, stuck with me, causing me to always have in the back of my mind that what I did (or didn't) do could result in death and destruction.

I could have stayed at DP&L, probably could have risen in the ranks after it went through the several evolutions of this company and its sister companies, but I actually found work often too plodding and interspersed with gaps to fill, which did not satisfy my hunger. So, I decided to move back into the consulting engineering sector, finding a job as a key EE with a very small firm in Dallas.

One milestone accomplishment while at the power company, was getting my registration as a professional engineer (P.E.). I was very proud to achieve that on a timely basis, which gave me the credentials to further my income and became a cornerstone of my entire career.

Chapter Three – "Back to Buildings"

The firm I joined was indeed small, and had three principal owners, two about my age and one somewhat older. All had matured working for larger firms in the Dallas area, doing large high rises, hotels, and other types of facilities. The primary owner was an EE, and he hired me to basically get him out of having to do the detailed electrical design work. Interestingly, the firm did more than just "MEP" design on architectural projects: they had established themselves as a recognized leader in "value engineering" studies, such as for the Corps of Engineers; and, they had developed their own system and protocol for managing the electrical energy cost and billing for large regional shopping center developers/owners. I saw both of these non-design aspects as beneficial to a small firm from a financial stability standpoint, as I was aware of the ups and downs of the building market. I do think it is always a good idea for any engineer considering a move to not just focus on the prospective job description, but to investigate the new employer's staying power in its market sector, and its strategic flexibility in pursuing other endeavors and markets.

As it turned out, this firm was what I needed at the time: lot's of work, long hours, final authority on my projects, and variety. Here is where I really began to learn about the design of lighting systems for the interior spaces of buildings, as well as outdoor lighting. Previously I had just "copied" from existing projects, but now I really dug into the basics and theory of lighting and illuminations systems and different applications, and became familiar with the myriad of luminaire (lighting fixture) generic types and the lamp types and technology (fluorescent types versus incandescent, et cetera). I actually enjoyed lighting design as an esthetic activity, as I would work very closely with the architects or interior designers to provide the function and ambiance of their mind's eye. One of my projects even one a small contest for its lighting design.

We did several large regional shopping malls, one in Baytown that I stopped by in 2013, surprised in a way to see it had gone the way of so many large malls, from originally having anchor stores and chain outlets, to being basically a "mom and pop" hosting facility for imports, and other goods and wares. When I designed the electrical for that system around 1978, I even wrote a magazine article about how I designed the several utility services to the building. This project also was unique in that the developer wished to perform "master metering" of electrical energy, buying energy from the utility through a utility revenue meter, then reselling the energy to tenants. At the time this was prohibited, and the developer chose to fight the State rule – and won. This changed the game from then on in Texas. As I recall I wrote a brief for use by the attorney on the proposed design. This project also gave me insight as to how some contractors handle their construction cost. We were actually working directly for the electrical contractor as this was what is called a "design-build" project, so I got to know the project manager well enough for him to show me his little "black book" literally carried in his shirt pocket. In it he would keep tabs of change order cost that were disapproved by the owner, and push the unapproved difference cost into the next large change. There is a cartoon that has been around for a long while, that shows a yacht with a dinghy in tow: the dinghy has a name of "Contract" and the yacht has the name "Change Order". I have found there is a great deal of truth in this cartoon.

Having already mentioned that this firm was involved in leading value engineering studies, primarily for federal projects, I had the opportunity to become the president of the North Texas Chapter of the Society of Value Engineering (SAVE). In this capacity my duties were to find a guest speaker for the monthly activities, and coordinate the venue and meals. I learned all I could about VE methodologies, and participated in a few studies. I liked the "brain storming" aspects of such studies, where no idea was berated, but considered in the context of increasing the value of a project or a project element. Value in this case is defined as function divided by cost; therefore, to increase value you can improve functionality (more bang for the buck), or seek to reduce the cost while not changing the function (or increasing it). This methodology became ingrained in my "design mind" and served me well on many occasions in the future: I embraced the concept as fundamental to how I approached any design.

After about one and a half years at this firm, the principal owner informed me they wanted to make me a minor principal, and awarded me with about five percent of the stock, and to boot, gave me a company car, a nice one, that frankly I was able to use at will for personal use (guess they had a good CPA!). We moved to a new location that was built to our needs, and life was rosy. Then it happened: we did some project that had a small computer room as I recall, and I began to learn about "computer rooms", "computer centers (called data centers today), and something about computer power and air conditioning. This seemed to be the field that pricked my imagination at that time, and I saw it as an exciting area to be in, but I did not see my firm then really pursuing such design as a specialty. So, I began to look around, and found another ad that would take me to the next phase of my career.

Chapter Four – "Computers"

A small company had started up recently, specializing in designing and even building computer rooms and centers, as well as a spectrum of other related consulting offerings, located in Arlington, Texas. I will mention the firm's name, as it is long ago "deceased": Total Assets Protection, Inc. So I buzzed over to Arlington one day to interview. The firm was obviously a startup, having taken up its offices in a totally non-functional cheap office building, but I really "dug" the entrepreneurial spirit of the principals and the staff of about ten that were already on board seemed really sharp. I got a nice offer, took it, and so began my submersion in the world of computers!

I would be the lead professional engineer, managing the design, working directly under one of the three principals. I also found that I would be able to engage in the construction of the data centers, and soon was specifying computer room air conditioning equipment, designing these, as well as security systems, environmental monitoring systems, and being the one to do a lot of the actual installation of the security and monitoring, meaning actually installing the cables and connecting the devices such as card readers, water detection take, et cetera. I also designed the fire detection and suppression systems with the assistance of favored local vendors.

The work proved to be a real, big cup of tea for my engineering interests, and I loved it. Moreover, as it turned out, we also did a lot of pre-design studies of existing computer centers, across the U.S. and even into Canada, giving me the opportunity to "strut my stuff" as a computer center "expert". So often once or twice or three times a month I was traveling with one or two associates to NYC, Chicago, DC, the west coast, and Canada, doing on-site surveys and then writing reports of our findings with recommendations on improvements. A small percentage of these would result in our being awarded a design-build contract, and this is where the "real money" was to be had.

This firm had been started as a debt-leveraged idea, funded by a single oil investor who had been convinced of a good return on his investment. The goal of the principals was to take the company public, and after I had been there about two years they were able to do just that. Suddenly the firm had a very lucrative influx from its IPO, and the principals decided to move us into two floors of a mid-rise building, into offices built to our own design. I was promoted to VP of Engineering Design, and was able to lure two old friends to join the firm, one as lead architect and another as lead mechanical engineer. Soon others joined, and we had a very nice complement of capable design staff, and along about this same time, we began to use computer aided design.

Over the years with this firm I learned all I could about every aspect of computer center design, whether it was physical security, site selection, power reliability, lighting, and all other system elements, and the pros/cons of each for given applications and locations. I carried that knowledge with me, adding it to the quiver of "smart arrows" I carried on my back from thenceforth.

As one of the managers, I also got to enjoy a few of the benefits, such as a management retreat with my wife to New Orleans, which as a fun experience. The principals also enjoyed golf so there was an annual golf festival, and although I did was not a player, I did participate and actually hit the longest drive to my surprise, and also kept my wife from running over others as she drove our cart and not being an automobile driver, just seemed to not be quite in control at all times of the cart.

Since I was getting deeply into the theory and practical application of power reliability, I began to write magazine articles on various subjects, published in several trade magazines. This experience allowed me to exercise my penchant for writing, which I had always enjoyed. Over the next decade or two I continued to publish a few more articles, not just about computer power, but about such topics as variable frequency drives for motors, and such. The study and preparation for writing each article was like a self-study course on the particular subject, as I wanted to be sure when it went to print with my name on it, I could defend the content if challenged.

Chapter Five – "Bigger Things"

I had been with TAP nearing four years, and had enjoyed it. My design group had successfully completed the designs on a number of centers, and had a capstone project under construction in the Las Colinas area of Irving, Texas: MTech, which was the computing operation of the then MBank, which had introduced ATM's. I was feeling like I had mastered the design of computer centers housing large systems produced by IBM, the giant of that day, and frankly was thinking of what the next step should be in my career. Once again I began to look around, and as had happened before, an add in the Dallas Morning News caught my attention. A large engineering firm in Albuquerque, New Mexico was looking to bring on board a new manager of electrical engineering. I applied, and was asked to come out, so my wife and I flew out for a night and day. I interviewed with the top manager, and learned this firm was indeed one of the larger national engineering firms, and that the Albuquerque office was one branch office, specializing mostly in federal projects, particularly at such locations as Los Alamos, Sandia, and other DOE facilities. They also were engaged in a lot of military base design work, as well as other governmental projects, the local airport and numerous and diverse projects. I learned that this larger company had just bought out a smaller local electrical firm in order to capture a greater local non-DOE market, and the combined electrical staff was about seventy-five, which I would manage. This sounded like just what I was looking for to advance my career, so I took the offer, and we prepared to move, finding a very nice pueblo style condominium to lease on a hillside in Coralles, overlooking the Rio Grande valley and Sandia mountains. We were located just down the road from where the annual hot air balloon festival takes place, which made for some great viewing of the magnificent balloons as they would float even low over our residence.

Well, this part of my career taught me that things are not always what they seem, and that I best be aware not to look through rose colored glasses in the future.

As soon as I was situated in my new office, next door to the aged founder of the firm that had been bought out, I discovered my first task was to begin laying off some twenty-five of my electrical staff, as directed by upper management. I soon learned the reason for this was that the purchase of the smaller firm had not properly reconciled the value of the existing contracts, and frankly there were serious problems with overruns.

So I immediately had my hands full: dealing with cost issues, getting to know my large staff and the leads capabilities and personalities, trying to win their respect, participating in securing new profitable work, as well as tacking a myriad of issues with new and existing projects in various stages of completion. In a few weeks I was sort of getting a comfort level and understanding of the lay of the land so to speak, and had been invited to have dinner with my boss and his wife (my wife had stayed in Dallas due to her career). I learned quite a bit about the firm at that affair, including that my boss actually was in some hot water over an assignment he had in Europe previously. As luck would have it, within a few weeks he opted to retire from the firm, and all of a sudden I was without my friendly "sponsor". Then the back biting began, and soon thereafter a new manager was brought in from another office, and in a few months my relationship with him soured. The reasons for this are not appropriate for this book, so I will just say that every work situation comes with the challenges of human relations, values and prejudices. The young engineer should be aware of his or her compatibility with the "norms" of the workplace, which often are latent and unseen on the surface.

I found most of the projects we were doing interesting, and usually they were one of a kind projects. I got involved in some pretty intriguing ones, including taking a lead role in proposing on a "Star Wars" weapon prototype that was to be installed at White Sands, a "maser beam" weapon that would be used – if it worked – to kill Soviet warheads and satellites (I never was clear on the actual intended purpose). I took the lead working with a local generator company and utility companies, to conceive and propose a scheme to fulfill the Federal procurement for a 750 megawatt power plant that could start up and deliver power to the test site in a matter of a few short minutes. We worked on the idea for several weeks, and I hosted technical meetings with key players, only to soon learn the "plug had been pulled" on Reagan's budget, so we were left holding a "cool" idea only.

The variety of work we did was considerable, including small health care clinics and hospital projects for the Indian Health System, mostly on the Navaho reservation. I took one long driving trip to Standing Rock, met one of the tribal managers and we drove out to a couple of small villages. This made for a very interesting day for me, just chatting with this smart and personable native American, my first chance to do so, although I had several Navahos and Pueblo Indians on my staff.

We also had projects at the DOE nuclear weapons facility, Pantex, outside of Amarillo, and I also had staff that worked at Sandia and commuted to Tonapah, Nevada weekly, and although I had to sign their time cards, I was not privy to what they actually did, since it was top secret/need-to-know assignments. A few years later I would be involved with other projects at Pantex with another firm.

My time with this Albuquerque firm was actually a good step. Although the day to day relationships with peers and office manager were lacking after I lost my "sponsor", the experience gathered in managing a large group of technical people and large, diverse and complex projects, each unique in nature, was invaluable for my future. So after about two years I began looking into how I could get back to Dallas, as I had been commuting back and forth by plane and car at least twice monthly, to my home and sweet wife still there. Again fortune smiled, and I responded to an ad placed by an equally large engineering firm, and soon left the beautiful state of New Mexico, which I had thoroughly enjoyed sightseeing.

Before leaving this part of my career review, I should also mention a couple of projects that I could not seem to shake free of after departing. One had to do with a project we had done for NASA, a space shuttle tracking relay station at White Sands. I had reviewed the "submittal data" of the major electrical equipment proposed by the contractor, and had rejected the submittal for not meeting the design specifications. I thought no more of it, until a year after being back in Dallas I was contacted by NASA, saying they wanted to depose me for a lawsuit the contractor had filed against NASA for delaying the construction due to the switchgear submittal rejection. So, on one nice morning, I was summoned to meet with a representative of NASA and the contractor, and the deposing para-legal, and was asked a number of questions, to which I responded, of course, honestly and roundly. I made it clear why I had rejected the submittal, and that to do otherwise would have diminished the project and not have given the installation less than what was paid for by NASA. After that meeting I never heard more about the case.

Another project actually necessitated my returning to Albuquerque over a weekend, as a private consultant, in order to assist with the final "cut-over" of a special programmable control system for Mountain Bell's very large emergency generator system in downtown. I had actually done the detailed design of this system myself, and was solely knowledgeable of the intricacies. So at approximately 6 p.m. on a Saturday, we were given the green light to begin testing the system, but our "window" for testing ended at 6 a.m. Sunday. To make the story very short, the first test failed, then the second, then the third, and I lost count. At each failure I was looking for what the gremlins might be. I was also sweating beads believe me, being on the "hot seat" in a big way. We were literally about 30 minutes from the end of our window, when I got this flash idea, out of the blue, that we needed to adjust one of the time delays in the controller by something like a half-second. The technician made the programming adjustment, and we ran the test once again, and it worked. After a long, tense night, we cheerfully went to breakfast, then I headed back to the airport. Slam dunk!

Chapter Six – "Career Peaking"

Returning to Dallas I took what was essentially a position that was equivalent to what I had left in New Mexico, except the number of staff was about a third, and I actually was department manager of both electrical and controls departments. Fortunately the small controls group had a lead engineer who really knew his business, as I had only a modicum of knowledge about control systems details, devices, and how these were typically presented on drawings and in specifications manuals.

Over the five years I spent with this firm, I had a generally great relationship with the other managers and for the most part with the guy in the corner office. I also was able to interact with the other branch offices' electrical managers, most who had "grown up" in the firm, and I learned to "lean on" them as needed for insight and knowledge, another lesson which served me well later on working with other multi-office firms.

We did a lot of unique, mostly one of a kind projects, ranging from military base projects to designing sophisticated industrial plants for semi-conductor manufacturing ("wafer fabs"), to electric furnace steel, and rolling mills, and cogeneration plants. As mentioned previously, I also got directly involved with projects at Pantex, and we did a couple of memorable projects there, including a new personnel security portal system (to enhance security of workers entering and leaving the top secret plant), and a new blast proof, tornado proof facility where weapons would be disassembled (as a result of the SALT warhead reduction by the US and the Soviet Union). The steel mills were fun, although as the manager I was often not directly designing the details, but performing guidance, oversight and quality assurance. I only had the chance to go out to a couple of operating steel mills (although I had worked at Armco Steel in Houston as a college student). Being up close and observing a 115MW arc furnace "do its thing", especially as an engineer who normally is thinking of preventing short circuits, is impressive. The sound is like a thousand locomotives passing by within a few feet!

Of all the project work we did, I got very personally involved in the design of cogeneration projects, due to my background at DP&L. "Cogeneration" is the term given to the process of generating electricity and capturing the waste heat of that system for use in some other thermal process, such as heating water for use in a laundry, as an example. During my last couple of years at this firm, we undertook getting into this market niche, and assemble a small team of a project manager, myself and two outside consultants, one who was adept at doing feasibility studies of potential cogeneration for clients that had existing heat and electrical power requirements, and another consultant who I brought on board, who I had worked with at DP&L, a very senior protective relay expert. We were able to land several feasibility studies and three of those panned out to our doing the design of the cogeneration plants. None of these were very large, ranging from 800kW to about 4MW electrical output, but each was unique in its own way. The small installation was actually a first for the State of Texas, located at one of the institutions in Austin. Another was for a regional hospital in Mississippi, which at the time of this writing I believe is still in use, some 24 years later. In fact, this installation provides emergency power to the hospital, and during hurricane Katrina, kept the hospital up and running while all other area hospitals faltered. When I read this a few years ago I was filled with satisfaction knowing I had a role in this project.

In 1993 there was a strategic repositioning within the firm, and one day I was called into the VP's office, and told that although I had the best record of managing electrical departments in the entire company, I was being let go so he could bring back a company veteran who was well versed in steel mills, a niche that he wanted to pursue more heavily. (Sadly, the US steel industry's days were numbered even then.) Soon I found myself capitalizing on the wafer fab experience I had garnered, and found employment at Texas Instruments as a consultant.

My two years at TI was again an assignment that allowed me to capitalize on my diverse knowledge. I spent a lot of time on power reliability issues, but also learned a great deal about wafer fabs up close and personal, as opposed to my prior arm's length design experience. I also found myself managing the final part of a major electrical construction project, which replaced about twenty medium voltage secondary substations with newer "double-ended" substations and made use of a scheme of "static switches" to enhance maintenance operations. The semi-conductor business has a long history of ups and downs, and I was released, which again provided an opportunity.

Chapter Seven – "The Dot.com Era"

I was able to very soon join a Dallas MEP design consulting firm, that had been through a high growth peak then diminished in size and underwent an ownership change, to a firm of about twelve people when I joined. At one time this firm had done some large high rises in downtown Dallas, and had a long and respected history helping build north Texas for three decades. I was asked to take over as the lead EE, and given the size of the firm I was directly performing much of the design on larger projects. As my good luck would have it, not long after I joined the firm, we were caught up in what has been called the "dot.com era", the explosive growth of the internet and telecommunications, build out of fiber optic cable in every city across the nation, call centers, and "help desks". At the time this was music to my ears, as we had been doing mostly corporate headquarters and large Class A office buildings around the Dallas area, all still in full operation nearly twenty years since, and I am proud to know these projects will continue to "do what buildings do" for many decades to come. With my background in computer centers from the early 1980's, I was a natural fit to take the lead in these newer forms of computer installations, such as what are called colocation sites. I will leave the reader to look up the term of "colocation data centers", as to really give an adequate treatment of the myriad of types and the service businesses these are built for is far beyond this present discussion. Suffice it to say, these all had (have) a need for backup/emergency power, usually by means of large diesel generators located securely within the perimeter of the data center, special redundant cooling systems and power systems, often designed as fully redundant and furthermore designed to allow concomitant maintenance operations without impacting the reliability to supply the critical information systems, and all the other aspects that we years earlier had addressed in designing "big box" computer centers.

During this period we also got involved not only with data centers, but with network communications sites, which I was not directly experienced in, but found I could bootstrap what I did know and quickly "mastered" the design of these facilities, often housed in larger "warehousing" facilities dedicated to hosting multiple telecommunications providers. At the time of this writing in 2018, a number of the various sites I provided the electrical engineering far back in the first few years of the twenty-first century are still operating and many have expanded considerably.

When the "dot.com" era begin to slow some, coupled with a management decision to pear down the staff to focus more on less high technology projects, I was advised that I was being released from the firm. Frankly, although it was a bit of a shock, I quickly realized that this might very well be my opportunity to go into business on my own. So I started a one-man engineering consulting business, under the DBA, "Power Vector Company". I also sold my homestead in Dallas, moved into a nice loft residence in which I could set up my office, had cards printed with my own logo, and commenced to "pound the pavement" looking for design engagements. I quickly found a few, data center related, which with my experience, was what I sought to specialize in. I did two or three projects modifying existing data centers, then all of a sudden my work in that market just seemed to dry up. I have my theory of why, but it is overly personal to burden this book. To stay afloat, as with most consulting firms, I had to loosen up my business model and take on much more mundane work, and found that in doing contract work for another small engineering firm. As it turned out, that firm, also a one-man contractor, had its own problems, and when they did not pay me a $5000 fee (small by any standard) that I needed to survive a few more weeks, I had to roll up my carpet and wait things out. So for a year I enjoyed living with my God children's mother and "the kids", all but one still in public schools, and became a soccer dad and housekeeper for a solid year, all the while trying to re-establish myself professionally. Finally after a year, with my finances beyond desperate, I secured a good position with a small, but well established firm in Austin, where I had began my career so many years before. So, it was off to help "keep Austin weird"!

Chapter Eight – "Architects are Fun People!"

The next nearly seven years of my career was spent in Austin, where I was often the project manager of MEP design on various architectural projects, often for governmental or higher educational projects around Texas, mostly though in Austin. I was intensely involved in the actual design of many such projects, all highly architectural and unique. On nearly every project we did, I worked very closely with the lead for the architectural firm, listening to their desires and wishes and of course, their client's needs and criteria. For my efforts I can point to a number of beautiful and functional buildings around Austin, and even to the "far reaches" of north Texas and the Texas panhandle that will be their long after I am gone, serving present or future purposes.

One aspect of the designs during this period was that various energy conservation codes and certifications began to enter and thoroughly permeate the building design requirements, some by law and some as voluntary goals of the owners. Austin during that period as a city was one of the leaders in the nation in adopting higher energy conservation requirements, and many owners were intent on also reaching higher, adopting the US Green Building Certifications, which encompasses far reaching requirements. It was in this time that such innovative design requirements as providing charging stations for electric hybrid automobiles began to be included. Also the controls for interior building illumination systems and outdoor lighting began to get quite complicated, as did the selection of lighting fixtures and lamps that would provide adequate lighting while meeting the "power density" (measured as watts/square foot, or similarly). These requirements added a level of complexity and design time requirements that actually made design work often less profitable, solely due the hours necessary to perform such services. This represented just one more turn of the screw that continues and seemingly will continue, pressuring design firms to work more smartly and use more efficient technology: a precarious situation, especially for smaller firms. I fear also that the quality of design work has already been impacted due to shortcuts, using less qualified designers who have less real understanding of design and the "real world" of the buildings they design, but these same younger designers have grown up with a mastery of computer aided design and application of various software to the design process. We may already be right on the border of artificial intelligence (AI) entering the design process, soon to be exponential in its "take over". Humans will soon have to re-evaluate their place in designing the built environment, I am certain.

Chapter Nine – "Ins and Outs of Hospitals"

As I mentioned the Austin firm I worked for was a sole proprietorship, fairly low wages and pretty much a sweat shop. As I neared seven years with the firm, one day I was called to my younger supervisor's office…..the proverbial corner office……and told I was being let go, as they could not even make the next payroll. I was shocked, but having been "there" before, was able to accept the news and process it positively. For a few months I searched for a new situation, biding my time and using it to rethink my life, take up meditation, increase my walks and workouts, dine and do a little site seeing around Austin. Then along came what turned out to be a very good opportunity, with a very nice salary increase compared to the prior: I would be the sole senior EE for the startup of a new office of this multi-office firm in San Antonio. First, however, until the San Antonio office lease was signed, I was to work out of the Houston office, which I did for several months, and I was given a nice corporate apartment to reside in during the week, and for a number of weekends I commuted back to my Austin abode awaiting the final move to San Antonio.

This firm was one of the major players in the hospital design business, and allowed me to reach way, way back in my career to the time I was designing the expansion to the old Austin hospital, and couple it with my understanding of emergency power and data centers, but also levy me with the immediate task to really dig into learning the codes pertaining to hospital design, and the many and complex functional areas within hospitals and clinics.

Hospitals represent one of the highest practical manifestations high-technology applications, whether its medical equipment, or the actual building itself with many sophisticated power, heating/cooling, medical gas, and other systems buried out of sight from patients and families. Also, every two or three years there was some revision to some code that the designer had to keep up with and apply. I enjoyed doing healthcare design, and it was made more "fun" for me in that I also got involved with the large central plants that provide the power, cooling and heating. One of my pet projects was the design of a large plant for a major children's hospital in Fort Worth, Texas. Working with the generator and switchgear suppliers, I developed a highly complex control sequence and logic for the power system, which gave me both nightmares and joyful highs over the course of the design.

After three years with this firm, I began to get uneasy about the apparent slow-down in our work load, and having been burned more than once, I decided to jump ship and take a position with a competing firm, that also did mostly healthcare design work in Dallas. I had high expectations that I could help turn-around and grow the Texas office, and was pumped up after visiting the home office in Florida. As it turned out, that was a mistake, the office had a history of talent and management issues, and after six months spent largely cleaning up the messes of predecessors and working on low-margin projects, I again bailed out, deciding to parley my design experience with my industrial design and operation experiences of long ago. Some may look at my voluntary departure from this firm, following a similar departure only a short time before, as an indication that I was not willing to make a commitment. I disagree entirely. Life is too short as it is, and at my advanced age by this time, I saw no reason to punish myself literally for the misdeeds of predecessors. I knew I could do better, and felt that I had a right to exercise my "perks of diverse experience", so to speak. I did, and joined my next firm again with a very positive outlook.

Chapter Ten – "Perks of Diverse Experience"

I accepted a position as lead EE with a Spanish-owned "EPC" (engineering/design/construction) firm located in Dallas, that specialized in various projects spanning several industries such as petrochemical, material handling and water treatment plants. I thought this would be a great opportunity for me to integrate and put to good use a large part of my knowledge repertoire. I was also getting in at pretty much the "re-tooling" of this firm that had recently gone through some ownership and structural changes. As it turned out, the firm was very poorly managed, and my immediate supervisor was totally out of his element – plus we were on polar opposites of social and political core values - and within weeks I just did not see any good future with the firm. I left, not having a clue what would be next in my career, and feeling disappointed by now, but comfortable with my decision and absolutely feeling that something better would materialize. Within weeks several key engineers and designers left that firm, and several months later that office shut down. So I did not regret my decision in any way.

I knew the data center market was jumping, as I had monitored that sector from a distance for the last few years, just to keep up with changes in technology and business in case I did one day need to get back into that game. That paid off, and I found a firm in Dallas that was indeed looking for a senior EE with data center skills. I was happy about this new job, and it had the big perk of just being a ten minute walk from my loft in downtown Dallas. I was enjoying my involvement in this office, a small contingent of both engineers and architects, and the team members, all decades younger than myself, were respectful and embraced me as just "one of the team". As it turned out, I had been brought on board as the firm had expected to be successful in acquiring some large project work for the Dallas office, and my resume was to help that effort. However, the proposals were not successful, and after a few months once again I was given a gentle send off, being told if work picked up I would be the first to be called back. I never was called back.

Chapter Eleven – "The Big Finish: Oops!"

By this time I was around seventy years old, and by most standards should have just retired. However, I was still in very good health, and intended to keep working "till I dropped", as I would often joke. Moreover, for financial reasons I thought postponing retirement a few years would be beneficial. So, I kept searching for that next great job, determined to keep very positive and believing something good awaited. Well, it did, and then it did not.

After several months of "sabbatical", I was brought on board by a very large international engineering firm, as part of the engineering and construction management team for George Bush Intercontinental Airport, to have oversight of the proposed new international terminal, and other electrical work supporting that element. My friend and I moved to Houston, found a cute cottage near downtown, and I began the daily commute that I had hoped to get away from, having gotten spoiled walking to work on the last job in Dallas.

I was part of a team of very, very talented and experienced engineers, project managers, cost estimators, scheduling experts and airport planners, and frankly was very impressed with the knowledge all these brought to the program. In a few months I had garnered the respect of team members, supervisors and key infrastructure managers of the Houston airport system. The challenge we were given was multi-faceted, and my own assignments included providing direct and detailed design input in support of two major elements of the program: providing new electrical service to the entire airport central terminal area, and providing the basic design concept for a tri-generation plant to be located at the airport. So I had my arms and brain around not only renovation/new construction of a major passenger terminal electrical systems, but also the power delivery to that, connected directly to the serving electrical utility. In theory this should have made me as happy as "a pig in slop". I failed to understand for some time, that this program had some serious challenges in planning versus proposed program cost, not to mention the politics that can and did surround such a large public project. So rather than moving ahead to stay on the planned schedule, traction was lost as weeks and then month after month the back and forth between "us" and the architect/contractors, stirred by political fingers, wore on and on. It did not take a rocket scientist to foresee what was to come: one day after being there for over eighteen months, ten of us were called into a small conference room, and informed we were being cut. I had seen it clearly coming, so there was not surprise, but I felt so sorry for some of the younger team members who had families and/or had bought homes believing they would be in the area for several years. Moreover, I kind of felt sorry for those who had to stay on in such a dismal work situation. Such is life in the built environment world of today.

Epilogue

We stayed on in Houston through the two year term of our lease, which gave me a lot of time to mull over my future. I am now fully retired, living back in Dallas. I am actually enjoying not being in the race any longer, free of the pressures of performance and the vicissitudes of project life. Life is a journey, as is often said, and I can fully appreciate the journey I have been on, and still look forward to.

Life is here to be lived!

ENJOY YOURS!

www.ingramcontent.com/pod-product-compliance
Lightning Source LLC
Chambersburg PA
CBHW030520220526
45464CB00006B/2883